D1567362

New Frontiers of Space

Cutting-Edge
Astronaut
Training

Karen Latchana Kenney

Lerner Publications ◆ Minneapolis

Lerner Publications Company
An imprint of Lerner Publishing Group, Inc.
241 First Avenue North
Minneapolis, MN 55401 USA

For reading levels and more information, look up this title at www.lernerbooks.com.

Main body text set in Adrianna Regular.
Typeface provided by Chank.

Library of Congress Cataloging-in-Publication Data

Names: Kenney, Karen Latchana, author.
Title: Cutting-edge astronaut training / Karen Latchana Kenney.
Description: Minneapolis : Lerner Publications, [2020] | Series: Searchlight books : new frontiers of space | Audience: Ages 8–11. | Audience: Grades 4–6. | Includes bibliographical references and index.
Identifiers: LCCN 2018054301 (print) | LCCN 2018058298 (ebook) | ISBN 9781541556706 (eb pdf) | ISBN 9781541555808 (lb : alk. paper) | ISBN 9781541574823 (pb : alk. paper)
Subjects: LCSH: Astronauts—Training of—Juvenile literature. | Astronautics—Juvenile literature. | Space sciences—Juvenile literature.
Classification: LCC TL855 (ebook) | LCC TL855 .K46 2020 (print) | DDC 629.450068/3—dc23

LC record available at https://lccn.loc.gov/2018054301

Manufactured in the United States of America
1-46035-43358-4/1/2019

Contents

Chapter 1

GETTING READY FOR SPACE

At the Neutral Buoyancy Laboratory in Houston, Texas, a crane lifts two people into a special pool. Each person wears a space suit that weighs almost 300 pounds (136 kg). Divers move the people to large metal structures. Underwater cameras watch the people closely. What is happening? The people in the space suits are training to be astronauts.

Astronauts are lowered into a pool at the Neutral Buoyancy Laboratory.

ASTRONAUTS TRAIN IN THE NEUTRAL BUOYANCY LAB TO PREPARE FOR FUTURE SPACE WALKS.

▼

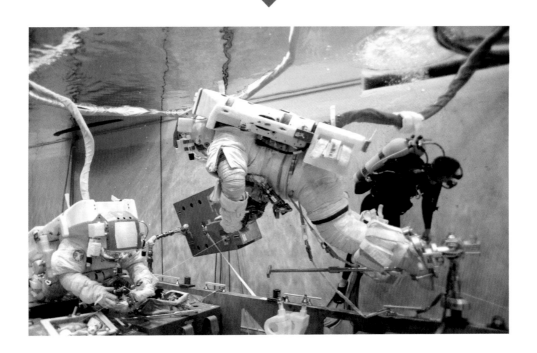

Floating underwater feels a lot like floating in space. It's one way astronauts train for an environment without gravity. It's part of their two years of training at the National Aeronautics and Space Administration (NASA) Johnson Space Center.

Project Mercury

The very first US astronauts trained at the Wright Air Development Center in Ohio in 1959. Thirty-one men went through physical and mental tests. The tests created conditions similar to what humans might go through in space. Machines spun and vibrated the astronauts. Loud noises and high heat tested the men's endurance. They learned how to control a spinning, out-of-control spacecraft. They practiced getting out of a space capsule in water. They learned how to survive in extreme conditions.

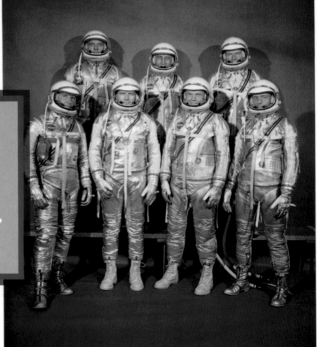

The seven Mercury astronauts were (*left to right, top row*) Alan Shepard, Gus Grisson, Gordon Cooper and (*bottom row*) Wally Schirra, Deke Slayton, John Glenn, and Scott Carpenter.

Seven men were chosen for the Mercury missions. The goal was to send a man into space, have him orbit Earth, and return safely to the surface. On May 5, 1961, Alan B. Shepard became the first US astronaut in space. His spacecraft orbited Earth for fifteen minutes and twenty-eight seconds.

Alan Shepard's space capsule was called the *Freedom 7.*

The G Machine

To escape Earth's gravity, a rocket must accelerate quickly. This can be hard on an astronaut's body. To train the first astronauts, NASA created the G machine. An astronaut sat at the end of a 50-foot-long (15 m) arm. The machine whipped him around in a circle, creating strong g-forces on the astronaut. G-forces measure the amount of force a body experiences when it accelerates. People pass out from too much g-force. This is because the heart can't pump enough blood to the brain.

Astronauts Today

Since that first spaceflight, astronaut training has changed a lot. Astronauts have gone from riding in the G machine to using virtual reality technology. Today's training prepares astronauts to go to the International Space Station (ISS) for months at a time. They learn how to maintain the station and use its technology. Astronauts also do science experiments. Today's astronauts have to be pilots, engineers, and scientists. Their high-tech gear and tools let them live and work in space and safely return to Earth.

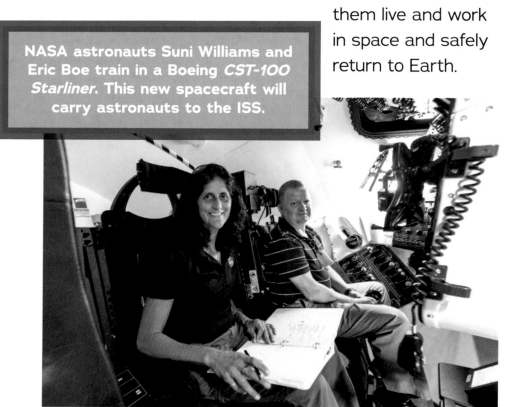

NASA astronauts Suni Williams and Eric Boe train in a Boeing *CST-100 Starliner*. This new spacecraft will carry astronauts to the ISS.

Chapter 2

SPACE GEAR AND TECH

One important piece of gear for astronauts is the space suit they use for space walks. It's like a mini-spacecraft that protects astronauts from space. The suits provide oxygen and keep the air temperature inside comfortable. Radios in the suits allow astronauts to talk with people in capsules or inside the ISS. Space suits have fourteen layers and take forty-five minutes to put on. Those layers make the suits stiff and difficult to move in. NASA is working on making a new, more flexible space suit. Then astronauts will be able to move around more easily and do more difficult jobs.

The pieces of a space suit lock together to completely cover the astronaut's body and protect it from the dangers of space.

Space Fact or Fiction?

The latest NASA space suit has a built-in toilet. That's a fact! NASA engineers are designing a space suit that has a waste disposal system. This system acts like a built-in toilet. Astronauts will be able to wear Orion Crew Survival Systems Suits for up to six days without taking them off. That could save the lives of astronauts during an emergency inside their spacecraft.

Virtual Space Walks

To learn what it's like to be in space, astronauts train in NASA's Virtual Reality Laboratory (VRL) in Houston, Texas. The astronauts wear a special headset and visor over their eyes. The equipment creates a 3D image of the ISS. Wearing data gloves, astronauts can move their hands and see them move in virtual space. They can go on a virtual space walk and move along the outside of the ISS. The VRL helps astronauts understand what it looks like to work on the station during a space walk.

Screens show astronaut Nick Hague's progress as he trains at the VRL.

The SAFER unit on this astronaut's back can help him return to the ISS in an emergency.

The VRL also prepares astronauts for emergencies. They practice using their SAFER units. Shaped like backpacks, these units have small jets that fire. If astronauts float away from the ISS, these jets can propel them back to the station.

Every astronaut since 1980 has trained in the SVMF.

Practice for the ISS

To feel what it's like in space, astronauts train in the Space Vehicle Mockup Facility (SVMF) in Houston. The large room contains life-size copies of sections of the ISS. It also has copies of capsules and other vehicles astronauts use.

The SVMF lets astronauts practice tasks that they will need to perform in space. They get comfortable using the technology and vehicles. They try solving problems too, such as fixing a broken computer panel similar to a panel on the ISS. After hours of practice in the SVMF, astronauts are ready for almost any situation.

Astronauts complete about two hundred different training sessions in the SVMF.

Chapter 3

TOUGH TRAINING

Being in space requires more than knowledge. Astronauts also need physical strength and endurance. G-forces make launches hard on a person's body. Living without gravity can cause problems too. Since everything floats, astronauts don't use their muscles as much as they do on Earth. Muscles can weaken and shrink. To help make up for muscle that they'll lose in space, astronauts get into top physical shape before launch.

Astronauts must be mentally and physically prepared to live and work in a zero-gravity environment.

Astronauts must pass physical tests before going to space. They must have perfect eyesight or use glasses to correct vision problems. They need to have healthy blood pressure. They must be between 5 feet 2 inches and 6 feet 3 inches (157 and 191 cm) tall to fit comfortably inside the spacecraft. Because much of their training is underwater, astronauts must also be strong swimmers.

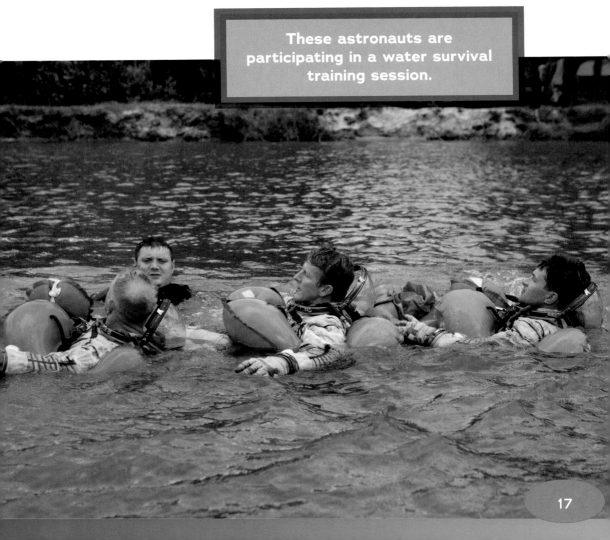

These astronauts are participating in a water survival training session.

From Pool to Space

To prepare astronauts for space walks outside the ISS,
NASA trains them in its Neutral Buoyancy Laboratory at
the Johnson Space Center. The laboratory is a supersize
pool. It's 202 feet (62 m) long, 102 feet (31 m) wide, and
40 feet (12 m) deep. Inside the pool are copies of parts of
the ISS, including its robotic arms.

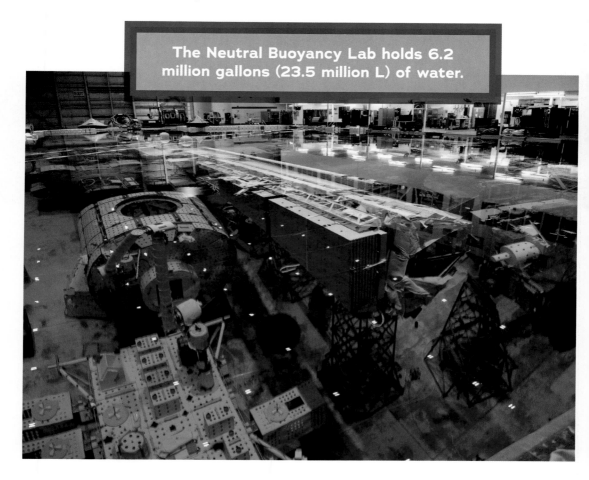

The Neutral Buoyancy Lab holds 6.2
million gallons (23.5 million L) of water.

Divers adjust the weights and floats on astronauts in the Neutral Buoyancy Lab.

The astronauts wear their heavy space suits in the water. Then divers put different floats and weights on the suits until the astronauts are the right weight to hover in the water. The astronauts practice doing different tasks on the ISS. They figure out how to move in their suits while weightless.

ASTRONAUTS EXPERIENCE WEIGHTLESSNESS DURING A PARABOLIC FLIGHT.

▼

Flight Training

Another way to feel weightless is to fly on a special aircraft. It can create short moments that feel like being in a zero-gravity environment. The plane flies up and then dives at a steep angle. This is a parabolic arc. As the plane dives, people inside feel twenty to twenty-five seconds of weightlessness. The plane has padded walls to protect astronauts while they train for future spaceflights.

The astronauts also fly a T-38 jet plane to train. This superfast plane helps them experience intense g-forces. G-forces are strong during a space launch. Astronauts get used to the feeling in a T-38.

NASA's T-38 trainer aircraft gives astronauts real-life experience flying at fast speeds.

Chapter 4

FUTURE ASTRONAUTS

Tomorrow's astronauts will travel to new, uncharted places. They'll go on longer trips in space and set foot on another planet for the first time. They'll also take regular people into space on space tours. On many of these tours, passengers would orbit Earth. Some people hope to take a trip around the moon.

SpaceX plans to send astronauts to Mars on a *Crew Dragon* spacecraft by 2024.

To prepare, NASA is training astronauts to work with private space companies such as SpaceX. The company's sleek *Crew Dragon* spacecraft seats seven astronauts. Boeing's *CST-100 Starliner* also seats seven passengers. Both capsules will bring NASA astronauts to and from the ISS starting in 2019. Someday the capsules will also bring tourists into Earth's orbit.

The first flight in the *Crew Dragon* spacecraft with astronauts on board is scheduled for June 2019.

To the Moon and Mars

NASA plans to build a space station to orbit the moon in the mid-2020s. NASA wants to study how living so far away from Earth affects humans. Astronauts will live there for one to three months at a time.

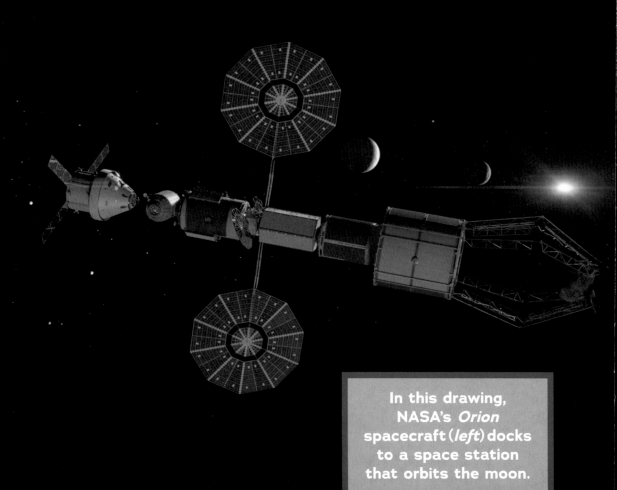

In this drawing, NASA's *Orion* spacecraft (*left*) docks to a space station that orbits the moon.

Scientists use X-rays to learn about the makeup of rocks on Earth. Astronauts will use a similar tool on Mars.

If the astronauts do well on the moon, NASA will use what they learned to send astronauts to Mars in the mid-2030s on the *Orion* spacecraft. It will be the farthest any person has traveled in space. To explore the planet, NASA is creating new tools astronauts can use. A laser device will scan and map the land. An instrument that uses X-rays will tell astronauts which minerals the rocks contain.

Working with Robots

As astronauts explore space, they'll use robots to help them. One of the newest robots is Astrobee. This flying robot can do many small tasks that astronauts do now, such as tracking thousands of tools on the ISS. Then astronauts will have more time to do other work. CIMON, a robot with artificial intelligence, can work and react like a human. CIMON could help astronauts do experiments and work through problems in space.

CIMON can recognize faces and even make eye contact while speaking to astronauts.

3D Printing

If you need something in space, it could take months to get it from Earth. Astronauts on the ISS use 3D printers to create useful objects such as tools more quickly. If astronauts become ill on a deep space mission, 3D printers could save their lives by printing the medicine they need. One day, 3D printers may create structures on the moon or Mars using dust that is already there. Astronauts may even print their own food!

It takes a lot of training and hard work to become an astronaut. Astronauts learn about the high-tech tools they need. They train their bodies to combat the effects of being in space. Most people never leave Earth. The few astronauts who go to space help us learn more about our tiny part of the universe. What will they discover next?

Astronauts will keep exploring and traveling farther into space than ever before.

3D Printer Activity

In 2014, astronaut Barry Wilmore was on the ISS and needed a wrench. NASA emailed him a 3D design file so he could print one. It was the first 3D object designed on Earth, emailed to an astronaut, and printed in space. If you have access to a 3D printer, follow the link below to download a design file. It's the same one that Wilmore used in space to print the wrench.

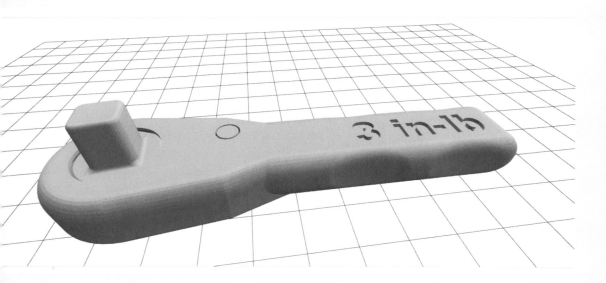

http://qrs.lernerbooks.com/dv00

Glossary

accelerate: to move faster

blood pressure: a measure of how easy or difficult it is for the heart to pump blood through a person's body

capsule: a vehicle that carries people and scientific instruments in space and can return to Earth

endurance: the ability to stay strong during a difficult experience

gravity: a force that attracts objects to one another

International Space Station: a spacecraft that orbits Earth where astronauts do experiments to learn about living in space

laboratory: a place used for scientific study and testing

parabolic arc: a u-shaped curve that planes fly in to produce moments of weightlessness for passengers

space walk: a mission that requires an astronaut in a space suit to go outside a spacecraft

3D printer: a machine that makes objects by laying down lines of material in stacks

virtual: something made by software that looks real on-screen

Learn More about Astronaut Training

Books

Goldstein, Margaret J. *Astronauts: A Space Discovery Guide*. Minneapolis: Lerner Publications, 2017.
Check out this book to learn more about how astronauts train and work in space.

Kelly, Scott. *Endurance: My Year in Space and How I Got There*. New York: Crown Books for Young Readers, 2018.
Learn about Astronaut Scott Kelly and his year living aboard the ISS.

Tunby, Benjamin. *Surviving a Space Disaster: Apollo 13*. Minneapolis: Lerner Publications, 2019.
Discover how astronauts survived an explosion on the Apollo 13 mission and came back to Earth.

Websites

ESA Kids: Astronaut Training
https://www.esa.int/esaKIDSen/SEM3RIWJD1E_LifeinSpace_0.html
Find out what astronauts from the European Space Agency do to prepare for space.

Our Universe for Kids: Astronauts
https://www.ouruniverseforkids.com/astronauts/
Learn about some of the first astronauts.

Smithsonian National Air and Space Museum: Astronaut Victor Glover on the Challenges of NASA Training
https://airandspace.si.edu/stories/editorial/astronaut-victor-glover
-nasa-training
Watch videos on this site about astronaut Victor Glover and astronaut training.

Index

Photo Acknowledgments

Image credits: U.S. Air Force photo by J.M. Eddins Jr., p. 4; NASA JBL/NBL, p. 5; NASA, pp. 6, 7, 10, 13, 16, 24, 27; Science History Images/Alamy Stock Photo, p. 8; NASA/Norah Moran, p. 9; NASA/Radislav Sinyak, p. 11; US Air Force photo by J.M. Eddins Jr, p. 12; Kumar Sriskandan/Alamy Stock Photo, p. 14; NASA/Josh Valcarcel, p. 15; GCTC, p. 17; ESA/Corvaja, p. 18; NASA/NBL, p. 19; ESA/Anneke Le Floch, p. 20; NASA/Jim Ross, p. 21; courtesy of SpaceX, p. 22; NASA/Bill Ingalls, p. 23; Kevin Lewis/Johns Hopkins University/NASA, p. 25; ESA/NASA, p. 26; NASA/Robert Markowitz, p. 28; Made In Space/NASA/MSFC, p. 29.

Cover: U.S. Air Force photo by J.M. Eddins Jr.